Groundwater Conflict 1970–

THE BITTER CONFLICT between Owens Valley ranchers and the Los Angeles Department of Water and Power ended in the 1930s, as described in the previous chapter. Many ranchers and merchants, seeing no future for their valley, moved away. Those remaining eventually realized that, although they had lost control of their water, they had not lost their dramatic mountain scenery and that a new future lay in promoting their valley and its mountains as prime vacation land.

During the next forty years, DWP continued to extend and enlarge its aqueduct system. By 1940 the system extended an additional 105 miles north, capable of gathering the waters of the Sierra Nevada tributary to Mono Lake as well as all waters tributary to the Owens River. By 1970 construction of a second aqueduct enlarged the system's capacity for export by 50 percent. To fill both aqueducts, DWP diverted four of the five streams flowing into Mono Lake and initiated large-scale pumping from wells in Owens Valley. As the consequences of pumping became evident, forty years of serenity ended abruptly. In 1972 Inyo County filed suit against Los Angeles.

Yesterday's conflict revolved around the right to use *surface water,* the right to divert running streams. Today's conflict centers on the right to use *groundwater,* which lies below the surface of even the most arid valleys. Does the owner of land have exclusive control over use of the water below? Does he have the right to drill and pump without limit—regardless of harm to plants, animals, air and people? An even broader argument questions whether anyone has the right to use *any* water without regard for the impact that use will have on other people and on the environment.

These are among the issues involved in the lawsuit filed by Inyo County against the Los Angeles Department of Water and Power. In the two chapters following, the special counsel to the County of Inyo and the chief engineer of water works and assistant manager, Los Angeles DWP, interpret the critical issues as they see them. Today's battle has been waged in the courtroom, and Inyo has looked to the law—not dynamite—for relief. The final outcome of this suit will have far-reaching implications for water use throughout the arid West.

June 1978   GENNY SMITH

Water for the Valley

HOW MUCH MORE OWENS VAL-
ley water shall go to Los Angeles?
This is the question fundamental to
one of the most significant environmental disputes of our time. The history of this dispute—of legal attack
and counter-attack—is an exciting story in itself. The ultimate significance of *Inyo* v. *Los Angeles,* however, goes beyond the immediate struggle
between county and city. For what we witness today, within our Deepest
Valley, foreshadows even larger and more difficult controversies in the coming
years: the efforts of regions with massive populations and immense appetites
for material and energy, to extract the last resources of regions with small
populations.

What right does one group of people, concentrated at the social and
productive "center" of society, have to use up the resources of the sparsely
populated regions? What right do the people in those regions, few in number,
have to keep what nature has given them—not just to protect their environment, but to preserve their own options for future growth and development?
These are questions that we all will face in our lifetimes on this planet.

1970: LOS ANGELES COMPLETES SECOND AQUEDUCT; STATE ENACTS ENVIRONMENTAL QUALITY ACT (CEQA)

This story begins in 1970. Two landmark events occurred that year, appar-
ently separate at the time, entwined ever since. The Los Angeles Department
of Water and Power (DWP or department) completed construction of its
second aqueduct from the Owens Valley to Los Angeles, enlarging by 50
percent its capacity to export Owens Valley water to the city. To fill
this "second barrel," the city for the first time implemented a systematic
program of extracting the valley's groundwater. But just as DWP was
putting its second aqueduct into operation, the state legislature enacted the
California Environmental Quality Act (CEQA), in response to public demands that government agencies pay heed to environmental values when
making decisions. As a starting point, the act required each agency to prepare an environmental impact report (EIR) on any project it proposed that
might have a significant effect on the environment.

Antonio Rossmann, author of this chapter, is special counsel to the County
of Inyo and one of California's leading public interest attorneys. Following
his graduation with honors from the Harvard Law School in 1971, he served
as law clerk to Justice Mathew Tobriner of the California Supreme Court.
In 1975 he became California's first ombudsman when he was appointed
Public Adviser to the California Energy Commission.

Los Angeles DWP increases groundwater pumping in Owens Valley.— Los Angeles, in deciding to build the second aqueduct in 1963, originally visualized a moderate program of groundwater extraction to supplement natural runoff from the Sierra in dry years. Based on its projected needs and on a simultaneous commitment to provide Inyo County ranchers who lease DWP land with a firm supply of water in dry as well as wet years, the department stated its intention to pump in an average year 89 cubic feet per second (cfs) of Owens Valley groundwater.* At the time the department announced this intention, approximately 31,000 acres of its Owens Valley holdings were irrigated and leased to Owens Valley ranchers.

Beginning in 1970, however, DWP constantly increased both its planned extraction of groundwater, and its actual pumping. In 1970 the actual pumping exceeded 90 cfs. Two years later, pumping exceeded 200 cfs. Subsequently the department proposed an average extraction rate of 180 cfs, with a dry year maximum of 376 cfs.

Effects of more pumping.—Valley residents, whether aware or not of DWP's plans, soon became aware of their effect. By 1972 two years of heavy pumping had dried up the valley's most popular and ecologically significant springs. The artesian wells along Mazourka Canyon Road, from which Independence townfolk had for years taken fresh water, stopped flowing; and the vibrant plant and animal community at Little Black Rock Springs was destroyed. This loss was not compensated for by DWP's creation of an adjacent artificial habitat at the discharge of one of its largest pumps.

The people of Inyo also noticed a dramatic increase in the frequency and intensity of dust storms during the windy winter months. This exacerbated dust level, with its attendant discomfort and aggravation of respiratory conditions in older valley residents, seemed to be generated by DWP's groundwater pumping. Prior to the pumping, the relatively high water table in the valley supported water-loving plants which gave color and protection to the valley's desert soils. But as the pumping drew down the water table, many of these plants died off, leaving the soil susceptible to wind erosion.

In an effort to placate valley concerns and to show that its pumping program would benefit the people of Inyo as well as Los Angeles, DWP in

---

* A cubic foot per second (or second-foot) is a measure of water flow: one cubic foot of water flowing past a particular point each second. Pumping at a rate of one cfs for one year will produce a total volume of approximately 236,000,000 gallons, or 724 acre-feet. Equivalents for some of the pumping rates referred to in this chapter are as follows: 89 cfs equals 57,500,000 gallons per day; 315 cfs equals 203,000,000 gpd; 666 cfs equals 430,000,000 gpd. Flowing at full capacity, the aqueduct in one year can transport at least 157 billion gallons from the eastern Sierra to Los Angeles. To complicate matters, reliance on *average* extraction rates is generally not meaningful. In California water seldom runs off in average amounts, but more often in extremes. Thus an average rate of 89 cfs could anticipate no groundwater pumping at all in wet years and more than double the average rate in dry years—at the very time when pumping might cause the most environmental damage.

1972 circulated a draft water management plan describing the intentions and effects of the plan. Although DWP assured them that the increased pumping would bring a more reliable supply of water to the valley, valley residents found little comfort in those representations. They were quick to observe that whereas DWP now promised to irrigate 11,000 valley acres with a "firm" supply, DWP had actually irrigated three times that much acreage a decade before. Moreover, the 1972 report confirmed the city's intentions to increase the intensity and scope of groundwater extraction. New wells would be drilled, and the 89 cfs long-term rate projected in 1963 had grown to 180 cfs.

1972: INYO COUNTY SUES LOS ANGELES CLAIMING IRREPARABLE DAMAGE AND CLAIMING CEQA MANDATES AN ENVIRONMENTAL IMPACT REPORT

When confronted with similar unilateral decisions by the city in the past, Owens Valley people had appealed to the state legislature for assistance and, by forceful actions such as dynamiting the aqueduct in the 1920s, appealed to public sympathy. In 1972, however, the people of Inyo believed that the legislature had already provided the relief they needed by enacting CEQA. So for the first time in their history, Inyo people through their county government appealed to the courts to enforce the law for their protection.

On November 15, 1972, District Attorney Frank Fowles filed a lawsuit in the Inyo County Superior Court. In that action, entitled *County of Inyo* v. *Yorty* (Inyo County Superior Court No. 9365) the county claimed that the department's groundwater pumping project was producing an irreparable environmental impact on the Owens Valley and that the department had failed to prepare an EIR on that project. On behalf of the county Fowles demanded (1) that Los Angeles be enjoined from extracting any more groundwater from the Owens Valley; (2) that DWP be ordered to prepare an EIR; and (3) that the court retain jurisdiction over the county's claim to ensure that no groundwater pumping would take place that would cause environmental damage in the valley.

Narrowly, then, the county demanded an EIR. More broadly, however, the county asked the court to prohibit any groundwater pumping that would harm the environment. The former claim drew upon the letter of the law in CEQA; the act clearly required EIRs on new projects with potentially adverse effects. But the county's latter claim—that the law substantively prohibited environmental damage to the valley—was not clearly authorized by CEQA. Five years would pass before the county's basic plea—that its environment not be destroyed—matured as a claim that the courts would recognize.

Superior Court rules against Inyo.—In response to the county's lawsuit, Inyo County Superior Court Judge Verne Summers issued a temporary restraining order against any increased pumping. But Judge Summers never had the opportunity to decide whether that injunction should continue or to adjudicate the county's claim and the city's defenses. Less than two weeks after the suit was filed, Los Angeles invoked a provision of state law which enabled the city as defendant to demand that the case be tried in a separate county. Judge Summers granted the department's motion to change venue, and the parties agreed to remove the case to Sacramento County, a neutral location that was reasonably accessible to both sides.

In January 1973 the matter came before Sacramento Superior Court Judge William White (Sacramento County Superior Court No. 228928). The county sought to obtain a preliminary injunction that would remain in effect until the trial was completed. On January 19 Judge White denied that injunction, concluding that the city's principal defense was meritorious: because the second aqueduct was completed and placed into operation prior to the effective date of CEQA, no EIR could be required on either the aqueduct or on the pumping that was initiated to fill the aqueduct.

Court of Appeal assumes jurisdiction.—The county now faced a turning point. By order of the Superior Court, the groundwater pumping and its adverse effect on the valley would continue. While the county could proceed with a trial in this court, Judge White's preliminary ruling did not offer Inyo much promise that its claim—that an EIR was legally necessary—would prevail. So District Attorney Fowles took the best route open to protect the county's position. On January 26, 1973, he submitted a petition to the Third District Court of Appeal in Sacramento, asking the appellate court to order Judge White to halt pumping until completion of the trial in his court. Such an unusual petition would be granted by the appellate court only if it believed the county's claim to be substantial.

The court of appeal did not issue the injunction that the county requested. Instead on February 26 it took an action of far greater magnitude and significance: the appellate court elected to treat the county's request for an injunction as a claim for final relief on the merits, and boldly assumed the duty of adjudicating that claim on its own, without trial in the superior court. In so acting, the court of appeal followed a rarely invoked but well-established procedure that allowed it to assume original jurisdiction over a claim of great public importance, whose significant issues must be resolved as soon as possible.

Thus the court of appeal in Sacramento became the primary forum in which have been resolved the county's claims for relief from DWP's groundwater pumping in the Owens Valley. Over the next five years that court would write for itself, the people of Inyo, and the people of our state one of the most distinguished chapters in the history of judicial response to an intense public controversy. The court's eloquent wisdom, expressed first through its Justice Frank Richardson (now a justice of the California Supreme Court) and subsequently through Justice Leonard Friedman, not only balanced and adjudicated the claims and interests of Inyo and Los Angeles. It also charted a course for all Californians to heed in meeting human needs from diminishing and remote natural resources.

1973: COURT SUSTAINS INYO CLAIM THAT LOS ANGELES MUST PREPARE EIR ON PUMPING

Following the court's assumption of original jurisdiction, Inyo and Los Angeles in March 1973 submitted briefs restating and refining their positions. Inyo claimed that notwithstanding the completion of the aqueduct in June 1970, the city's continued and expanded extraction of groundwater since that time formed a "project" under CEQA for which an EIR was required. The county cited specific examples of destruction of the valley's environment, such as the drying of natural springs, to dramatize the need for an EIR. Los Angeles, on the other hand, argued that in 1970 its second aqueduct was already completed and, as early as 1963, was approved and financed on the premise that systematic groundwater extraction would take place in the valley. Thus CEQA should not now be retroactively applied to determine if that which was already approved should be reapproved.

On June 27, 1973, the court of appeal issued its decision. In that opinion (*County of Inyo* v. *Yorty* (1973) 32 Cal. App. 3d 795) the court sustained Inyo's claim. Noting that CEQA required decision makers in government to identify and evaluate environmental factors and to prevent irreparable harm before it is too late, the court emphasized that an EIR was at the core of this now-mandatory process. In order to carry out the law's intent, the court held, the completed project of the second aqueduct should be separated from the ongoing project of groundwater extraction. An EIR should be prepared on this ongoing project so that its potential to harm the environment would be revealed.

Three-year dispute over the rate of interim pumping, 89 cfs or 211 cfs.—The court then turned to the difficult task of determining the rate of groundwater pumping during preparation of the EIR. The court noted the county's claim of perceptible damage caused by pumping. At the same time, the court recognized DWP's vital role as supplier of water to the state's largest city and the city's need for a reliable and adequate source of water. Weighing these considerations, the court declared a temporary pumping rate of 89 cfs—the pumping rate in November 1970 when CEQA became applicable—and directed the Sacramento Superior Court to conduct further hearings to determine an interim pumping rate that represented an average of extraction for the wettest and driest years between 1970 and 1973.

Although the court of appeal may have anticipated that its orders would be carried out with dispatch, three full years were devoted to disputes over the interim pumping rate and over the scope of the EIR. When the superior court in October 1973 conducted its first proceeding to refine the pumping rate, DWP over the county's strenuous objection convinced the court to permit pumping up to 221 cfs on a fiscal year (July to June) average. The county brought its objections to the court of appeal; that court in a brief order of September 4, 1974, set aside the 221 cfs rate, re-established the 89 cfs rate temporarily, and ordered the superior court to attempt a more equitable application of the 1970–1973 average.

Three-year dispute over the scope of EIR: should it encompass only pumping for use in Owens Valley, or pumping for aqueduct export too?—In its preparation of the EIR, the department proceeded on the following premise: because the court of appeal required an EIR for the pumping program but not for the second aqueduct, groundwater pumping for aqueduct export was not part of the project; instead, the EIR would only evaluate the project of increased pumping for use in the Owens Valley. Thus, not only would the EIR address pumping of a much smaller scope, but also the *alternatives,* which CEQA required to be identified and considered, would only embrace alternatives of less water left in the Owens Valley. In a nutshell, Los Angeles proceeded with these strong convictions: that in 1963 the total export of water from the valley—from groundwater as well as surface sources—was planned to maintain a level of 666 cfs; that DWP had already approved a groundwater extraction plan to maintain that level; and therefore DWP need not evaluate in its EIR any change of that export level.

Dispute over irrigation supply for valley ranchers.—Until September of 1974 the county objected, with little efficacy, to what it viewed as a deliberately truncated definition of the groundwater pumping project. In that month an unfortunate decision by Los Angeles brought this issue to a dramatic head. As related above, on September 4 the court of appeal had set aside the superior court's 221 cfs rate and temporarily reinstalled the 89 cfs rate. In response, DWP announced on Friday September 20 that it would terminate the Owens Valley ranchers' irrigation supply the following Monday. In justifying this move, which the department's aqueduct engineer later characterized as "educational," the department claimed that it was forced to cut back because of the district attorney's success in rolling back the 221 cfs rate.

The department's precipitous action immediately brought the county's new district attorney, L. H. "Buck" Gibbons, back into Judge White's Sacramento courtroom. Gibbons sought and obtained on September 27 a temporary restraining order against cutting back irrigation supplies. He also sought to enjoin further processing of the EIR on grounds that it purposely misdefined the scope of the groundwater pumping project. Once the temporary restraining order was issued, however, the other matter never came to hearing. Instead, the department agreed to withdraw its draft EIR and prepare a revised draft and, in preparation of that draft, to consult systematically with the government and citizens of Inyo. The county and DWP also agreed to a total groundwater extraction of 68,000 acre-feet for the winter of 1974–75, to govern instead of the 89 cfs annual rate that the court of appeal installed, pending the superior court's resetting of the rate. In May 1975 the superior court reset the pumping rate at 178 cfs, again over the county's objection.

In its subsequent EIR preparation the DWP did establish formal mechanisms for receiving county input. Not only did it provide the county with advance copies of its draft, but it also established an EIR task force of Owens Valley citizens with whom it met regularly. Despite these gestures, however, Los Angeles never succeeded in overcoming the valley's general distrust for the process and substance of DWP's efforts. More often than not, it seemed to the valley folk, DWP was using the public meetings not to receive comments or criticism, but instead to sell its preconceived assumptions and judgments. Despite repeated valley objections to DWP's assumption that its EIR did not have to address groundwater for export, DWP insisted that the only project for which the court ordered an EIR was that of groundwater extraction for Owens Valley uses.

DWP certifies EIR and approves large-scale pumping program.—In May 1976 the controversy fully matured. Just as the briefing had been completed in the court of appeal on Inyo's claim that the superior court had erred again in setting an excessive pumping rate of 178 cfs, the department published its final EIR and announced its intention to certify it within the month. These actions meant that DWP would shortly approve the expanded groundwater pumping program. In response to the comments on its EIR, DWP had modified its program somewhat. Rather than a long-range average pumping rate of 180 cfs and a maximum of 376, DWP now proposed a long-range average rate of 140 cfs with a maximum of 315 cfs. No new wells were to be drilled. Nonetheless, aqueduct export remained fixed at 666 cfs.

In its three-volume report justifying these decisions, the department devoted but a handful of pages to discussion of the alternatives that meant most to Inyo county: conserving water in Los Angeles and obtaining more water from its other historic source, the Colorado River. Despite the county's strenuous objections, the city showed no intentions of altering either its EIR or its groundwater project. With no relief apparent to the county short of a major court battle, District Attorney Gibbons recommended to the Inyo County Board of Supervisors that it engage special counsel to meet the demands of litigation, and on June 1, 1976, the board adopted the recommendation.

Los Angeles' formal approval of the project and certification of the EIR were scheduled for June 3. The district attorney and special counsel obtained a brief postponement in order to address personal appeals to the Los Angeles Board of Water and Power Commissioners, Mayor Thomas Bradley and

City Attorney Burt Pines—asking each to exercise responsibility and avoid a legal fight that, in the county's view, the city could only lose. Bradley and Pines never responded, and the Water and Power Commissioners on July 15 certified the EIR and approved the full groundwater pumping project.

1976: DISPUTES OVER
ADEQUACY OF EIR AND INTERIM PUMPING RATE

Anticipating that DWP would certify the EIR and then ask the court of appeal to dismiss Inyo's suit because the EIR had been completed, the county had one week earlier seized the initiative in that court. Urging the court to reassert the original jurisdiction that it had assumed in 1973, the county asked the court not to discharge Los Angeles' responsibilities until it had evaluated the adequacy of the department's EIR. The law not only required an EIR, claimed Inyo, it required an adequate one. In addition, until such adequacy was determined, the court must continue to restrain groundwater pumping in the valley. Furthermore, the county asked the court of appeal to set the interim pumping rate itself. In response the city argued that, by completing its EIR, it had discharged its duty to the court and that no need remained for the court to establish pumping rates.

Court rules it will evaluate EIR.—At the extraordinary oral argument of July 21, 1976, the court of appeal resolved some of these claims and established the framework for resolving them all. Departing from its accustomed pattern of hearing formal argument for 30 minutes at most and of announcing its decision weeks or months later in a written opinion, the court in this instance devoted more than two hours to reach this immediate conclusion: that its 1973 mandate required not only an EIR, but an *adequate* EIR. Furthermore, that adequacy must be judicially reviewed; the court established a process for that evaluation.

Court sets interim rate at 149 cfs.—In its subsequent written opinion of August 17 (*County of Inyo* v. *City of Los Angeles* (1976) 61 Cal. App. 3d 91), the court of appeal reaffirmed its original jurisdiction over Inyo's claims and determined itself the pumping rate to govern pending review of the city's EIR. It accepted Inyo's claim that the superior court's rate was too high and that it should have been set on a water-runoff year (April to March) rather than a fiscal-year basis. Nonetheless the court rejected Inyo's claim that the court abandon the 1970–1973 average and constrain DWP to the 89-cfs rate that coincided with CEQA's effective date. Taking judicial notice of the dry condition prevailing, the court fixed the rate at 149 cfs, subject to the significant condition that DWP provide valley users their customary supply.

1976: INYO REVIVES THE MAJOR ISSUE:
DOES CEQA PROHIBIT HARMFUL PUMPING IF DWP
HAS A LESS DAMAGING ALTERNATIVE?

For Inyo, the opportunity was now ripe not only to void the EIR, but also to revive its original claim that the law under CEQA prohibited Los Angeles from unnecessarily harming the environment of Owens Valley. Thus in late 1976 Inyo urged the court (1) not only to reject the EIR because of its faulty project definition and failure to assess meaningful alternatives; (2) but also to reject the department's *decision* as a violation of the California Constitution's mandate that all the state's water be conserved; and (3) to enforce CEQA by ordering Los Angeles to reject its environmentally harmful pumping program and accept the less damaging alternative of water conservation in Los Angeles.

To these claims the city responded (1) that its EIR fulfilled the requirements of the court's 1973 order; (2) that its EIR incorporated the findings of more than fifty public meetings and consultations with valley officials and citizens; and (3) that the groundwater pumping would benefit the valley by providing it with a greater supply of water than would be possible from only surface supplies. As to Inyo's constitutional claim that Los Angeles conserve water, the department argued that Inyo county could not press such claim because it owned no water rights competing with those of Los Angeles; moreover, conservation in Los Angeles lacked relevance to a groundwater pumping program designed to benefit water users in the Owens Valley.

DROUGHT OF 1976–1977 INTENSIFIES THE CONFLICT

Before it could resolve these claims, however, the court of appeal became the vortex of even more intense conflict between county and city—conflict produced by the severe California drought of 1976–77. By its August 1976 order setting the rate of 149 cfs, the court of appeal seemed to force Los Angeles to react. During the relatively dry 1976 water year, when surface supplies in Owens Valley were low, the department filled the aqueduct by pumping at rates sometimes exceeding 200 cfs. Cut back in August to a 149-cfs rate, the city turned to its other sources of water—the Colorado River and the California Water Project.

But as 1976 ended with even less precipitation than 1975, amid predictions that 1977 could produce the worst drought in California's history, even southern California's vast water supply system became overtaxed. By year's end 1976, as reservoirs dwindled, most northern California communities had adopted water conservation measures saving 25 to 60 percent. Because the California Water Project could not meet all its municipal requirements and leave any significant supplies for Central Valley agriculture, the State

Department of Water Resources obtained an important agreement from the Metropolitan Water District (MWD). (MWD serves a large part of southern California and at times sells water to the DWP.) MWD agreed to forego most of its 1977 importation of northern California water and to rely exclusively on its Colorado River supply.

Nonetheless, officials in the southland contended that with only moderate ten percent conservation effort, all needs could be met. Los Angeles' problem, then, was not a shortage of water but a matter of economics. Colorado River water from MWD generally cost Los Angeles more than groundwater from the Owens Valley.

Los Angeles asks the court to allow pumping at a maximum rate of 315 cfs.— In late February 1977 the department cited these facts to the court in support of its motion to pump at a maximum rate of 315 cfs. In further support of its positions, the department submitted unsworn statements that it had solicited from Owens Valley ranchers, whom the department had led to believe would receive irrigation supplies only if the pumping rate was increased.

Pointing out that the drought had produced greater distress in Inyo than Los Angeles, the county responded by urging the court to reject the department's claim in light of Los Angeles' failure to adopt a single water conservation ordinance. The county also criticized the department for misrepresenting the ranchers' position and presented a sworn statement from the president of the Inyo Cattlemen's Association that the department had failed to disclose to the ranchers that the court's August 1976 order guaranteed them their customary supply of water.

In an extraordinary preliminary memo, court replies that until Los Angeles conserves water, its request to extract additional Owens Valley groundwater is not likely to be granted.—The court's response to these claims spoke to all Californians and to all times. Issuing a preliminary memorandum on March 24, but four days after hearing argument, the justices wrote that Los Angeles' failure to adopt an effective conservation program, standing alone, would compel denial of DWP's motion to increase the pumping rate:

> In relation to the state's current water crisis, the effort at voluntary conservation is inadequate to justify the requested relief. The California Constitution abjures the waste of water and seeks its conservation in the interest of the state's entire population. When the state's water resources dwindle, the constitutional demands grow more stringent and compelling, to the end that scarcity and personal sacrifice be shared as widely as possible among the state's inhabitants.

Moreover, the court did not find impressive Los Angeles' argument that it should increase its Owens Valley extraction in order to save the higher purchase cost of Colorado River water:

> Unless and until the municipal government of Los Angeles installs and implements methods which are predictably capable of achieving substantial water

savings and demonstrates a need for water rather than rate preservation, its motion for leave to extract additional underground water from Owens Valley is not likely to achieve success.

Although the court had before and has since published lengthy decisions sustaining the county's claim under CEQA, its brief four page memorandum of March 24, 1977, may likely endure as its most significant teaching. In the midst of the worst drought in our history, the court said that shortage required the sharing of resources throughout the state. Sharing became not a gesture but a *constitutional duty;* the state is but one ship in which all its citizens are equal passengers. If one region or city does not face hardships as great as those in other parts of the state, it must if possible restrict its own use of resources to alleviate the greater hardship of other Californians.

Equally important, the court's ruling suggested that before new resources could be extracted, the alternative of conservation must first be implemented—a highly significant rule that not only spoke to the conditions of the 1977 drought, but that also speaks to all future developers of natural resources. Before completing its review of the department's EIR, the court would further clarify this valuable rule—but not before having to face yet another critical conflict between county and city.

As noted above, in August 1976 the court ordered DWP not to reduce its supply of water to Owens Valley users below that customarily maintained since May 1975. In the valley's view, this order required the normal supply of irrigation water to begin on April 1 as customary. The department, on the other hand, in its public statements indicated that *if* allowed to pump at the maximum rate of 315 cfs, it would then "be able" to supply irrigation water at *half* the 1975 rate. April 1, 1977 passed, and neither county nor ranchers received assurances when the irrigation season would commence or how much water would be available. The next day, when the Board of Supervisors received the department's request for permission to increase pumping, the Board urged the department not to threaten contempt of court by failing to supply the normal amount of irrigation water. By April 11 the department had provided no response, except for a press statement that it was confused as to whether the court meant for water to be supplied at the 1975 or 1976 rate. In the valley, a department official stated that *no* irrigation water would be released without further order from the court. The county then asked the court to order irrigation forthwith. Only after this motion was filed, did the department announce that on April 15 it would begin supplying irrigation water—but at half the normal rate.

Again the court acted with unprecedented dispatch. The county's motion was filed on April 13, the department's response on the 21st. The court heard oral argument on Friday, April 22, and issued its decision the following Monday. It ordered the department at once to provide irrigation water during the 1977 season at 75 percent of the normal full supply.

Although the county had asked for the full supply in 1977—to compensate for the 50 percent cutback the prior year and the poor condition of the pasture lands—county and ranchers welcomed the relief that the court provided. The court's order of a 25 percent cutback in 1977 conformed to the water conservation goal that Governor Edmund G. Brown, Jr., had requested of all Californians and that the county had demanded of Los Angeles. Even more importantly, the court's action proved to the ranchers and other citizens of Inyo that the department would fail in its effort to play "water politics." For the moment the department was frustrated in its attempt to manipulate the irrigation supply as a dividing wedge between the ranchers and other citizens of Inyo.

1977: COURT RULINGS VINDICATE INYO'S CLAIMS

With the interim pumping rate and irrigation supply disputes momentarily resolved, county and city awaited the court's judgment on the lawfulness of the EIR and on the city's decision to expand its long-range groundwater extraction. On June 27 the answer came: virtually total vindication of Inyo's claims.

EIR must include pumping for aqueduct export.—By its decision of that date (*County of Inyo* v. *City of Los Angeles* (1977) 71 Cal. App. 3d 185) the court of appeal held that DWP's EIR was legally inadequate and refused to certify the department's compliance with CEQA. With forceful and precise language, the court traced the tortuous history of DWP's EIR preparation. Characterizing DWP's misinterpretation of the project to exclude groundwater for export as "serious," "wishful" and "egregious," the court pointed to its consequences. Not only had Los Angeles evaded an assessment of the project's impact on Inyo, but it had also concealed from the citizens of Los Angeles as well as Inyo the true nature of the groundwater pumping proposal and its impact on the people and environment of both communities.

CEQA requires Los Angeles to select the alternative least damaging to the environment.—Responding to the county's claim that DWP had failed to consider the constitutionally mandated alternative of water conservation in Los Angeles, the court wrote:

> The underlying policy and express provisions of CEQA limit the approving agency's power to authorize an environmentally harmful proposal when an economically feasible alternative is available. Notably, the Los Angeles EIR omits another alternative, one freighted with costs other than dollars. The omitted alternative is a tangible, foreseeably effective plan for achieving distinctly articulated water conservation goals within the Los Angeles service area. It is doubtful whether an EIR can fulfill CEQA's demands without proposing so obvious an alternative.

In this brief passage the court charted again a new course for county and city and all others to follow. Not only must conservation be examined as the alternative preferred by the California Constitution; but also the substantive provisions of CEQA, as interpreted by the legislature and Supreme Court, *required* selection of the conservation alternative if conservation (as contrasted to extraction) would result in less damage to the environment. Never before had a court so ruled.

The court concluded its opinion with an admonition to the city: it should not await the compulsion of further judicial decrees to fulfill its legal duties—duties not necessarily limited to preparing a valid EIR on Owens Valley groundwater pumping. In addition to the duties of conservation and of rejecting harmful projects in the face of less damaging alternatives, the court cited the advice of California Deputy Attorney General Larry King that DWP would most faithfully fulfill the law by preparing a comprehensive EIR on *all* of its water gathering activities, which would enable the city each year to select from its many sources the conservation and extraction pattern that minimized harm to the environment.

CITING DROUGHT AND DWP'S WATER CONSERVATION PROGRAM, COURT GRANTS CITY'S PETITION TO PUMP 315 CFS UNTIL MARCH 1978

The court's landmark ruling was but two days old, however, when the city boldly petitioned the court anew for permission to pump groundwater at a maximum rate of 315 cfs. This time the department came to court better prepared. In response to the court's March 24 memorandum, the city had instituted a mandatory conservation program calling for 10 percent reduction in water use; actual savings were in excess of 15 percent. Of at least equal significance, the city's motion was supported by the Metropolitan Water District. MWD claimed that its other consumers in southern California would be harmed if Los Angeles exercised its lawful right to purchase more MWD (Colorado River) water. Basically the city argued that since the drought was so severe that all water supplies and groundwater basins in the state were being drawn down to their limits, the Owens Valley groundwater basin should not be excepted.

The county objected to these arguments. Even though the city had for the first time in its history implemented a mandatory water conservation plan, its performance did not match that of northern California urban centers, or even the 25 percent cutback which the court had ordered in Owens Valley. Moreover, argued Inyo, MWD should not complain until it had achieved 25 percent savings in its entire service area. The county also stressed again that it had suffered from man-made as well as natural drought, because of the pumping since 1970. Finally, the county urged that the 149-cfs rate be maintained to provide DWP an incentive to complete an adequate EIR and comply with the law.

A tense courtroom heard these arguments on July 21. The justices, normally inquisitive and lively at oral argument, spoke little. But one day later their decision came: provided that DWP pump only from deep wells and that the city maintain 15 percent conservation, the pumping rate until March 1978 would be doubled to 315 cfs.

The city greeted this news with elation. City Attorney Pines in a press release hailed his efforts successfully protecting the city's rights; later he cited the July pumping order as proof that his office had achieved victory in its litigation against Inyo County. Other southern California communities also expressed relief; the City of San Diego put aside plans for implementing mandatory water conservation.

The people of Inyo were stunned. How could they have prevailed totally in their legal claims a month before, only now to have the fruits of that victory denied? Why should Los Angeles take all Owens Valley water, while saving only 15 percent, when Owens Valley was expected to and willing to save 25 percent? What purpose did it serve to challenge Los Angeles in court and win on the merits, if *in extremis* the city to the south and its neighbors could prove that "might still makes right?"

The county nonetheless took this defeat with dignity. According to some, Inyo could accept defeat easily because she had become conditioned by years of the city's feudal bondage. But other people in the valley—especially Inyo's Board of Supervisors—saw the need to do better and to work harder. The county engaged a professional hydrologist, to overcome the DWP's monopoly on expert knowledge of the groundwater supply, and then looked beyond the courtroom to find a long-term resolution of its dispute against the city.

STATE SUPREME COURT DENIES CITY'S PETITION
THAT IT REVIEW COURT OF APPEAL'S DECISIONS

However, the last and greatest judicial victory of 1977 belonged to Inyo alone. Frustrated in its efforts to implement a groundwater program free of the county's objection and judicial supervision, the city doubled its legal forces and in early August petitioned the California Supreme Court to review the court of appeal's decisions. Not only did DWP argue that the court should have accepted its EIR as adequate. It also claimed that *all* of the court's decisions since 1973—including its decision to exercise original juris-diction, its ruling that the city must prepare an EIR, and its restraints on groundwater pumping—had been in error. The city and its supporters, in-cluding the Metropolitan Water District and the *Los Angeles Times*, firmly believed that the Supreme Court would release them from the court of appeal, which in their view had become "an adversary contestant dueling with Los Angeles to restrict export from Owens Valley and to reallocate its water rights."

The county responded to these charges and then awaited the decision. If the Supreme Court accepted Los Angeles' petitions, Inyo's five-year effort to secure justice from the courts would be cast into doubt for the months or years that would be required for the Supreme Court to render a new decision of its own. If the court denied the petitions, however, Inyo's positions—and the propriety of the court of appeal's forward-looking mandates—would be vindicated.

On October 6 the word came, ironically from Los Angeles. While sitting in that city, the Supreme Court by unanimous vote and without elaboration denied Los Angeles' petitions. The next day's *Los Angeles Times* headlined, "Owens Valley Wins Major Water Battle over L.A."

WHAT LIES AHEAD FOR VALLEY AND CITY?

For the people of Inyo, and the people of Los Angeles, the Supreme Court's decision came at the right time. For even as the two were battling in the summer of 1977, the California Department of Water Resources and the Attorney General were bringing county and city to the conference table to explore the possibilities of cooperative management of the water resources in the Owens Valley for the mutual benefit of both parties. The county gratefully accepted this invitation, hoping to secure a voluntary plan that would produce not only an EIR, but also a water management plan that would protect its environment and still fairly provide both Los Angeles and Inyo with water. DWP, while also accepting the invitation of the state, withheld any commitments until the Supreme Court completed review of the case.

With the high court's final word received and with both sides standing to benefit from cooperation rather than confrontation, county and city have continued to meet. Many difficult problems remain. Hopefully they will be ameliorated as the county becomes a more active participant in the evaluation of hydrologic data and the decisions that flow from such evaluation. In the end, the success of this effort will depend upon both county and city accepting the wisdom that Justice Friedman imparted in the court of appeal's 1976 decision: "Neither party can have what it wants or needs; rather the needs of both must be recognized and balanced."

Nor should resolution of the groundwater dispute halt further cooperative efforts in the valley. Other problems of long standing must also be addressed, with all concerned recognizing not only the lawful prerogative of the city as owner of land and water rights, but also recognizing the need to restrain the absentee landlord's actual and potential abuse of power in preventing Inyo citizens from determining their own future. Among the problems today, in the valley's tourist-oriented economy, business people need more security than the five-year lease to which DWP currently restricts its town properties. The city will earn much good will when it forthrightly offers these properties for sale or longer lease, in a spirit of accommodating the existing economy rather than withholding one more implement of suzerainty. Similarly, ranch leases can and should be written to give each lessee security and a firm commitment of water, while still protecting the city from the creation of competing water rights or claims. Finally, in supplying domestic water to the towns, cooperation rather than conflict should prevail. If the department were to charge valley customers a fair rate for valley water, and if it recognized the need for independent Public Utilities Commission review of its valley rates, then the people of the valley would welcome additional measures to prevent waste of water.

During the past decade the Owens Valley has served as the battleground from which have emerged experiences and rules of law to which others—in the West, the nation, and the world—will look for guidance in the coming years of increasing shortages and sharpened conflicts. Let all who read this recent history recognize, however, that Inyo's greatest achievement and the court's greatest reward lie not in the protection of Inyo's inanimate resources, but rather in the renaissance of self-respect and self-determination in her *people*. Ahead of us waits the next and greater question: whether genuine cooperation—between the powerful and the few, between the urban center and the rural valley, between the consumer of vast resources and the dwellers on the land from which those resources come—will also emerge from America's Deepest Valley.

Water to the City

THE ISSUES RAISED BY INYO County's lawsuit against Los Angeles' pumping of groundwater from under its lands in Inyo County for beneficial use in Owens Valley and Los Angeles extend beyond environment, energy and economics.

GROUNDWATER IN THE DEEPEST VALLEY:
IS IT A QUESTION OF USER OR USE?

These broader issues can be focused by asking, "Is it a question of user or use?" The answer is based on our lifestyle preference, the money we are willing to commit to achieve that, and our willingness to make sacrifices. The answer will reflect society's view toward the development of renewable natural resources, such as water, and nonrenewable resources, such as oil.

A word about writing style: the arguments made against the city's activities in Owens Valley frequently appear in a style relying heavily on adjectives which color fact and create varying shades of meaning. That style has a strong emotional tone which is important to keep in mind.

OWNERSHIP OF WATER

One facet of the "user or use" question is ownership of water. The right to *use* water can be obtained by any person, agency, city, etc., if certain laws and administrative regulations are followed.

Los Angeles' rights to water in the Owens Valley were developed according to those laws and regulations. In 1905 the city posted a notice of its intention to use 1000 cubic feet per second (cfs) of Owens River in Los Angeles and filed a copy with the Inyo County Recorder. In 1934 the city applied to the State Water Rights Board for permission to divert 200 cfs from streams tributary to Mono Lake for use in Los Angeles. A permit was issued in 1940 and a license, which confirms the amount of reasonable beneficial use, was issued by the State Water Resources Control Board in 1974.

Rights to use groundwater are based on the ownership of property overlying a basin. When conflicts arise between pumpers, they settle among themselves or in court since the legislature has not established laws providing for the acquisition of rights through the permit process as it did for surface water.

Paul H. Lane, author of this chapter, is Chief Engineer of Water Works and Assistant Manager of the Los Angeles Department of Water and Power. Mr. Lane lived in Owens Valley in the early 1930s near Keeler where he attended school and from 1961 to 1966 near Big Pine. He was a member of the Big Pine School Board. His Department career began in 1949 and he has been involved with the Owens-Mono operations since 1961. He has a fond appreciation for Owens Valley.

The City of Los Angeles owns 240,000 acres in Inyo County, most overlying the Owens Valley groundwater basin. As a result of conflicts between pumpers in the Bishop area 50 years ago (the *Hillside* case), the city does not pump groundwater from a certain area in and around Bishop for export. There have been no other conflicts between pumpers in Owens Valley that have resulted in a limitation on the city's pumping for local use or export.

THE SECOND LOS ANGELES AQUEDUCT

Groundwater was one water source for the second aqueduct which was approved by Los Angeles in 1963 to increase export to 666 cfs. The other sources were to be streams in the Mono Basin and savings achieved by increased efficiency in irrigating the city's Owens Valley lands.

With the second aqueduct, the use of the Owens Valley groundwater basin was expanded. Historically, the city had relied upon the underground as a storage reservoir to maintain the supply to the city. Pumping rates as high as 188 cfs occurred during droughts in the early 1930s and 1960s. The new use to be made of

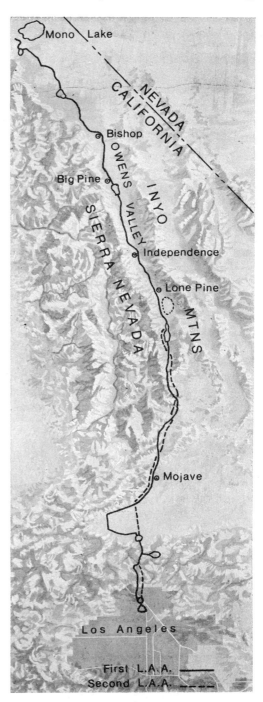

Los Angeles Aqueduct System

the basin was to salvage some of the water wasting to the atmosphere because of high groundwater levels. The 1968 Inyo County General Plan discusses the water salvage development potential in Owens Valley:

> By establishing a ground water operation it would be possible to control the levels of the ground water table in such a way as to prevent loss of water consumed by evaporation and transpiration. A mechanism could be established to maintain quantities of water required in conjunction with surface storage supplies to maintain a full flow of the Los Angeles aqueduct system; plus to maintain the water levels necessary to control evaporation and transpiration. (Page 25)

The concept of salvage relates to the *reasonable* beneficial use of water as distinct from beneficial use. Although the Supreme Court in *Hillside* found that subirrigation resulting from high groundwater levels was a beneficial use, the use was not *reasonable,* considering the facts of that case.

The projected pumping rates for the second aqueduct varied from a maximum of 250 cfs in the driest years to zero in the wettest. The average would be 89 cfs. Because the proposed pumping would be less than the average inflow to the groundwater basin (400 cfs), there would be no continual lowering of water levels as in the San Joaquin Valley, where pumping exceeds average inflow.

Need.—There were four reasons for building a second Los Angeles aqueduct. First, the Supreme Court in *Arizona* v. *California* issued a decision in 1963 that established the amounts of water that Arizona, California and Nevada could divert from the Colorado River. Ultimately, when Arizona completed an aqueduct to utilize its share, the Metropolitan Water District of Southern California (MWD) would lose approximately half of its flow in the Colorado River aqueduct. The Central Arizona Project is now under construction and should be completed by 1985. Parenthetically, MWD's rights are being jeopardized by another claim—that of the Navajo Indians.

Second, the city needed another aqueduct to beneficially use all the Mono Basin water permitted by the State Water Rights Board. If this water was not used, the city would lose part of its filing. Had that happened, water supplies to southern California would have had to be increased from the State Water Project.

The amount of water contracted for by MWD from the State Water Project is based on Los Angeles obtaining 666 cfs from the second aqueduct. Parenthetically, the State Water Project is also in jeopardy. Facilities in existence are sufficient to deliver one-half the water to which the state is obligated.

Third, the quality of Mono Basin streams and Owens Valley groundwater would be superior to that of the Colorado River and State Project waters.

Fourth, water from the second aqueduct was projected to cost $25 per acre-foot. Water from the MWD, which cost $29 at the time the second

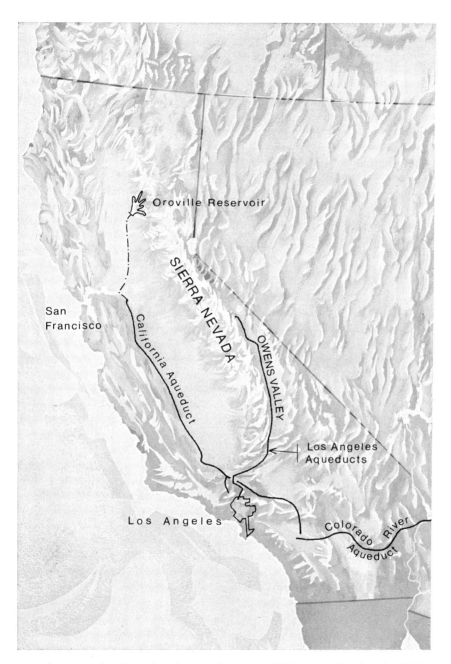

Aqueducts serving Los Angeles or Southern California: the State's California Aqueduct brings water from the Feather River; MWD's Colorado River Aqueduct from the Colorado River; and the Los Angeles Aqueduct from the Owens River and Mono Basin.

[221]

aqueduct was approved, was projected to cost $75 in the early 1970s. The rate effective July 1, 1978, is $95. Los Angeles Aqueduct water, excluding any credit for generation of electricity, cost less than $40 per acre-foot.

Water use in the Owens Valley.—Before the second aqueduct project, water for irrigation of 30,000 acres of the city's lands had been available on a feast or famine basis. When there was more surface water than needed to fill the aqueduct, city lands were irrigated. In dry years there was no irrigation and wells were turned on to maintain export to Los Angeles. Irrigation was cut off during five years between 1948 and 1970 and reduced during two.

A part of the second aqueduct project would replace the hit-or-miss irrigation method. The best 15,000 acres of the intermittently irrigated 30,000 acres were to be selected and supplied water year-in and year-out by pumping groundwater. Thus, even though the average acreage would be less, the firm supply combined with the higher productivity lands would result in net improvement for ranch lessees. No plans were made for use of water on city lands other than irrigation of 15,000 acres and use by livestock.

Construction.—The estimated cost of the second aqueduct was to be approximately $100 million. Because the city's pumping during the 1930s and 1960s was at rates near those planned with the Second Aqueduct, only two percent of the construction cost was for new wells. The aqueduct, begun in 1964, was completed and placed in service on June 26, 1970.

Inyo's study of second aqueduct.—Inyo County knew about the water sources and operation of the second aqueduct project. Their knowledge came from meetings with the city, a 1964 report by Stoddard and Karrer, consulting engineers who had been hired by Inyo to review the city's project, and a 1966 report by Los Angeles made in response to a resolution introduced by Senator William Symons (Inyo). This latter report described aqueduct system operations and local water use, both historically and forecasted with operation of the second aqueduct.

A WATER AND LAND USE PLAN

In September 1967, Inyo County petitioned the state to prepare a "comprehensive watershed protection plan" for the Owens River Basin and to prohibit increased export until the plan was adopted. Discussions on such a plan occurred off and on until November 1971, when representatives from the county, state and city agreed that DWP would prepare a water and land use plan. A detailed water study would be published first. That would be summarized and combined with the land use element to complete the second document. Public meetings were held in spring 1972 to obtain input. A draft of the water report was completed in October 1972 and given to the county and DWR (State Department of Water Resources) for comments. The projected maximum and average pumping rates in the report were 376 cfs and 147 cfs, respectively. The average rate represents a 65 percent increase over the rate approved in 1963 for the second aqueduct.

Why the increased pumping?—Several changes occurred between 1963 and October 1972 that required increased groundwater pumping. One thing didn't change—the average export to Los Angeles: it has always been 666 cfs.

What did change? The amount of land to be irrigated. In the negotiations with lessees to select 15,000 acres of the most productive land to be irrigated on a year-in and year-out basis using groundwater, DWP agreed to approximately 19,000 acres. The supply for this greater acreage would have to come from groundwater.

Also, DWP had planned to construct windmills, pipelines and troughs to supply water for livestock as part of the second aqueduct project. Those plans were not implemented. Instead, most stockwater continued to be diverted through unlined canals and ditches, a method of supply that required more groundwater pumping.

Another need resulted from DWP's participation in the Interagency Committee on Owens Valley Land and Wildlife (formed in 1970). The committee has established several recreation/wildlife areas that use water, such as the Buckley Ponds Wildlife Habitat Enhancement Project near Bishop.

Further, DFG was planning to expand operations at fish hatcheries and rearing ponds in the Owens Valley beyond the levels that could be supported by natural spring flow. Pumping would be necessary, and even though much of that water would flow through the hatcheries and contribute to the aqueduct system, the average pumping would increase because the wells would have to be operated during wet years.

THE EIR

Inyo County's November 1972 lawsuit seeking an EIR on increased pumping for export came as a surprise. Not only had the second aqueduct been in full operation before the California Environmental Quality Act (CEQA) took effect on November 23, 1970, but it had been operating for more than two years prior to the lawsuit.

DWP prepared the EIR mandated by the Third District Appellate Court's June 5, 1973 decision on a project defined as increasing pumping above the rate of 89 cfs for use in Owens Valley. This was based on the court's separation of increased pumping from the second aqueduct.

Consultants were retained. In January 1974 DWP formed the Owens Valley Groundwater EIR Advisory Committee of 17 residents of the Owens Valley having a diversity of interests. In mid-1974 Inyo County hired two consultants to assist them in review of the EIR's hydrology, flora and fauna. The draft EIR was completed in August 1974, one month after the date approved by the superior court in October 1973. A revised draft was published in January 1975 that included a discussion of alternatives calling for reduction in export from the Owens Valley.

After public meetings and technical workshops in Owens Valley, comments on the draft were evaluated. Consultants did additional work and more data was collected. The final EIR was released in May 1976. It contained detailed responses to comments made by members of the advisory committee and categorical responses to comments of others. The recommended project had been modified. Copies of the final EIR were given to the state for comments. None were received during the normal review period provided by CEQA's guidelines. The final EIR was approved by the Los Angeles Board of Water and Power Commissioners July 15, 1976.

Inyo objected to adequacy of the EIR, citing incorrect project definition and inadequate environmental assessments. On June 27, 1977, the appellate court found the project defined incorrectly, finding that the construction of the second aqueduct was separate from its operation and that all increased pumping above the long-term historic average should be considered rather than just the increment resulting after the passage of CEQA in 1970. About the city's environmental assessments, the court said:

> The project concept does not vitally affect the "impact" sections of the Report. The forecasts of environmental consequences in the Owens Valley are premised upon a long-term pumping rate of 140 cfs, which approximates the "project" as conceived in this Court's decision of June 1973. Thus, the informative quality of the EIR's environmental forecast is not affected by the ill-conceived initial project description.
>
> Inyo County strongly criticizes the environmental impact sections of the EIR, charging that the report understates the harm to flora and fauna of the Owens Valley and fails to describe air pollution potentialities. Courts are not equipped to select among the conflicting opinions of warring experts. It is not the function of the Court to determine the accuracy of the report's environmental forecasts. Reasonable foreseeability is enough.

On February 27, 1978, the court, in denying $85,000 of attorney fees to Inyo County's special counsel Mr. Rossmann, noted, "Its [Inyo County's] resistance to the environmental impact report was not impelled by the report's deficiencies but by its own litigational interests."

OWENS VALLEY ENVIRONMENT

A representative of the Sierra Club stated, "We recognize that Los Angeles is probably the savior of the Valley . . . our goal is to preserve the Valley as it is now." (*National Geographic*, January 1976, page 123). Living in Owens Valley is a life apart from the pollution, congestion and other urban problems familiar to us all.

Air quality.—Air quality is among the best in the state, with visibilities of 50 miles, 80 percent of the time. Infrequent dust storms, less than two per year based on visibility records at the Bishop airport from 1959 to 1975, arise from all parts of the valley. An article in the April 1876 Inyo newspaper notes such a storm.

The air basin implementation plan approved in 1971 states that a monitoring program for particulate will begin when funds become available. Not until 1978 did an approved sampling program begin, the last basin in California to start such a program. The delay in monitoring is testimony to the absence of a problem.

Other evidence came in 1976 when the Inyo County Planning Commission acted on Lake Minerals Corporation's proposal to expand by 8,000 acres the salt recovery operations on Owens Lake. Information from the China Lake Naval Weapons Center indicated that the salt recovery operations at the lake were the single most important source of dust during wind storms. The commission approved the expansion of the salt recovery operations with a negative declaration (no significant impact) based on a one-page initial study which did not mention dust.

Vegetation and groundwater.—Groundwater is within 10 feet of the floor of Owens Valley in most areas. As a result, water evaporates from the soil and plants. Evaporation from the soil is not beneficial. The use by plants may be beneficial. Whether it is a reasonable beneficial use depends on the value of the water and uses the water would otherwise serve. Groundwater pumped by DWP would be taken away from some plants and used by people and business in Los Angeles and for ranching, recreation and support of wildlife habitats in Owens Valley.

DWP's pumping is from lower zones in the groundwater basin. These are separated from the shallow zone, which the plants draw from, by zones of clay. The result is that fluctuations in the shallow zone caused by deep pumping take 5 to 10 years or more. The vegetation changes that ultimately would take place were described in the EIR.

DWP expanded its monitoring of water levels and vegetation in 1975. After three years, vegetation appears to be most affected by the amount and seasonal occurrence of rainfall and the patterns of livestock grazing. Shallow water levels declined two feet or less over most of the valley floor during that period.

REASONABLE WATER USE

Los Angeles.—The per capita use of water in Los Angeles was relatively stable from 1958 to 1978 at roughly 175 gallons per person per day (gpcd). Per capita use is all water used in the city (residential, commercial, industrial and governmental) divided by total city population. This figure of 175 gallons per capita per day is about average for the South Coastal Hydrologic Study Area, which extends from Ventura County to the Mexican border (DWR *Bulletin 198,* page 16). On a statewide basis, this per capita use is the lowest; the statewide study area average is 340 gpcd.

One reason for the city's relatively low per capita use is that the city began metering in 1903. People pay for all the water they use. Beginning in

December 1977, all residents started paying for water at the same rate, regardless of the amount of use. The same is true for commerce and industry.

The city has a voluntary conservation program which began before and continues after the mandatory provisions in effect during the 1977 drought. Kits to reduce water use in showers and toilets have been made available free to the entire city, and more than 600,000 kits have been given out. Hundreds of thousands of information brochures and educational materials have been given to citizens and schools. There is an active speakers bureau. Programs also assist industry, and DWP has crews assigned to detecting leaks. An ordinance prohibits hosing off sidewalks and driveways, untimely repair of leaks, serving water in restaurants unless requested, and the use of nonrecycling fountains.

Owens Valley.—Per capita use in Owens Valley ranged from more than 500 gpcd to more than 1,400 gpcd (DWR, *The California Drought 1977— An Update February 15, 1977,* page 149). This does not include any agricultural use. There were no meters in Owens Valley until 1976 and residential use is unmetered in 1978.

Irrigation is predominantly flood irrigation of uncultivated land; less than one-ninth of the acreage is sprinklered. The average use is four acre-feet per acre. In contrast, Water Code Section 1004 specifies that 2.5 feet per acre is considered beneficial for uncultivated land. Water is delivered through unlined canals and ditches; most other farming areas use pipes or concrete canals.

WATER RESOURCE DECISIONS

I think of water resource decisions in terms of the three Es: environment, energy and economics. Balance is important. The following is a list of primary considerations in each category for the use of groundwater from the Owens Valley basin.

Environment.—In Owens Valley, the issues are aesthetics, air quality, vegetation and wildlife. In Los Angeles, the issue is air quality because the oil that would have to be burned to produce the same amount of electricity as produced by water flowing through power plants along the aqueduct would add to air pollution in the Los Angeles area.

Energy.—Water from Owens Valley flows by gravity to Los Angeles and produces electricity at hydroelectric power plants enroute. An alternate supply from the Colorado River or State Aqueducts has to be moved over mountains using pumping plants that consume energy.

Economics.—In Owens Valley, economics is related to recreation and ranching enterprises that depend on water. In Los Angeles, economics is related to the cost of alternate supplies from the Colorado River and State Aqueducts.

SOMETHING OTHER THAN GROUNDWATER?

The October 10, 1977 *Wall Street Journal* discussed Owens Valley ground-water and Inyo's goal in terms of control of destiny. I have heard that issue repeatedly and I believe it has prolonged and exacerbated whatever real controversy has existed or exists between the city and Inyo over water, whether surface or groundwater.

IT IS A QUESTION OF USER

I believe the answer to the opening question is: it is a question of user. The competition for resources is such that it can be no other way. To allow a resource to be unused may be beneficial but not reasonable. The State Constitution mandates that the state's water resources be put to use to the fullest extent possible. To this end, the needs for water should be reasonable and the resource development program represent a balance between the three Es—environment, energy and economics. I believe there is such a balanced program for Owens Valley groundwater and that the city will preserve Owens Valley as one of California's scenic treasures because that goal is compatible with protection of the city's water supply.